T0029486

The Shell Spotter's Guide

Queen scallop (p. 68)

The Shell Spotter's Guide

HELEN
SCALES

Illustrated by
Ella Sienna

Published by National Trust Books
An imprint of HarperCollins Publishers,
1 London Bridge Street
London SE1 9GF
www.harpercollins.co.uk

HarperCollins Publishers,
Macken House,
39/40 Mayor Street Upper,
Dublin 1, D01 C9W8, Ireland

First published 2024

ISBN 978-0-00-864131-3
10 9 8 7 6 5 4 3 2 1

A catalogue record for this book is available from the British Library.

Printed and bound in India

If you would like to comment on any aspect of this book,
please contact us at the above address or national.trust@harpercollins.co.uk

National Trust publications are available at National Trust shops or online at nationaltrustbooks.co.uk

This book is produced from independently certified FSC™ paper to ensure responsible forest management.

For more information visit: www.harpercollins.co.uk/green

Contents

Introduction

Like many young marine biologists, I began my explorations of sea life with a bucket in hand down at the beach, searching for treasures in rock pools, among piles of seaweed scattered across the sand. Seashells were the first marine species I learnt to recognise and identify, starting with easier things such as limpets and mussels, and gradually moving on to trickier varieties that required a guidebook and some time looking at them closely. I learnt to tell apart the periwinkles from the top shells, and in time I got my eye in and spied the tiniest specimens hiding under rocks.

Shell spotting is a pastime that has stayed with me (although now I usually leave behind the bucket). Walking along a beach, I can't help looking down at my feet to see what the sea has brought in. On British shores, if I'm on my own or with patient companions, I'll spend time hunting for a few species that I've come to love the most – the ones I'm most likely to slip into my pocket and bring home with me. I have a special fondness for the cone-shaped pink and purple shells of painted top shells. And for me, there's no greater shell-spotting joy than searching through piles of shelly debris until my gaze finally falls on a single, perfect cowrie shell with its unmistakeable delicate corrugations.

Looking for seashells and learning their names is the perfect way to connect yourself to the ocean and to the living wonders of nature. They are the hardy remains of sea-dwelling animals, called molluscs, that either live down beneath the waves or reside between the tides – those ones you also have a chance to spot while they're still alive and crawling around or fixed firmly in place on rocks.

Seashells are the molluscs' external skeletons, or exoskeletons. These are protective, mobile homes that keep a mollusc's soft body safe inside. Molluscs make their shells from calcium carbonate, or limestone, which they secrete from a soft part of their body called the mantle. They obtain the component calcium and carbonate ions from seawater.

Molluscs often begin life with a tiny shell when they hatch from their eggs (many start life as tiny, swimming larvae and go through several rounds of metamorphosis, radically changing their body

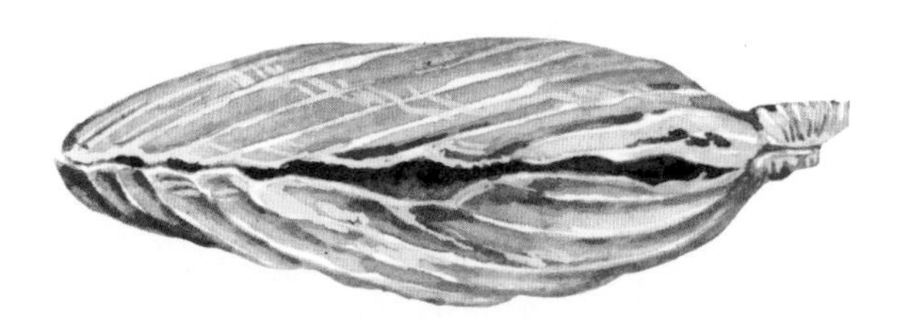

Queen scallop (p. 68)

shape before landing on their final adult form). Unlike crustaceans, such as crabs and lobsters, molluscs don't moult their exoskeletons but keep the same shells throughout their lives. As they grow bigger inside, molluscs gradually expand their shells by adding more calcium carbonate layers to the opening.

The majority of seashells that you'll find on beaches around the British Isles come from two groups of molluscs:

Gastropods (or snails): these molluscs usually make spiralling, coiling shells, except for limpets, which are snails that evolved uncoiled shells. Gastropods move about with a single, muscly foot (the word gastropod means 'stomach foot') aided by copious amounts of slime. Some gastropods seal the opening of their shells with a trapdoor, called the operculum. The central hole on the underside of many gastropod shells is called the umbilicus.

Many sea snails are carnivorous and sniff for the underwater scent of prey through a tube called the siphon. Others are herbivores and rasp at seaweeds with a rough tongue-like structure called the radula.

Most of them live in the sea, including various species that have lost their shells and are commonly known as sea slugs. There are also gastropods that evolved to live in lakes, rivers and ponds, and some moved all the way out onto land.

Bivalves: these molluscs make two shells that are often fan-shaped and finely ridged. You'll usually find single bivalve shells, such as cockles and mussels, lying on beaches, because they become disarticulated after the mollusc dies. Now and then you'll find an intact pair. In life, the twin shells are hinged together, with strong muscles to open and close them.

Bivalves are commonly filter feeders, sifting small particles of food, such as plankton, from the water around them. They draw seawater in and out of their bodies through a pair of siphon tubes.

There are some freshwater bivalves, but none are known to live permanently on land.

Less common mollusc varieties on our coasts include:

Cephalopods: including squid, cuttlefish and octopuses. Most have lost the ability to make external shells, although cuttlefish and some squid have internal shells. A rare group, known as argonauts or paper nautiluses, are the only octopuses that make shells, but you're very unlikely to see one of these shells washed up on a beach around the British Isles.

Chitons: molluscs with shells in eight plates arranged across their backs. They live exclusively in marine habitats, where they fix their foot to rocks.

Shell collecting etiquette

- It's okay to take home a few shells whose occupants are long gone (I don't recommend taking living or recently dead animals, unless you want a stinking mess at home a few days later). Don't take every last interesting shell you find; leave some for the next shell spotter who comes along.

- Try leaving only footprints and taking only photos. Photographs of living molluscs at the shoreline can be very useful to help you identify them later at home. You can also contribute photographs from your smartphone, tagged with a GPS location, to online biodiversity surveys such as iNaturalist.

- And, of course, be sure to write the details of your finds in this book to save them for the future.

Rocky shore explorations

- Always turn rocks back over when you have looked underneath. All sorts of animals and seaweeds are living on the rocks in conditions that suit them best, and they will suffer and may die if left exposed on an upturned rock.

- Some of the most interesting molluscs on rocky shores are only accessible at very low tides. Check tide tables and plan a visit for low spring tides.

- Ideal footwear for shell spotting on rocky shores is either wellington boots or neoprene dive/surf booties with a hard sole.

Staying safe on the shore

- Check tide times and, if you want to access the lowest parts of the shore, aim to arrive half an hour or so before low tide.

- Always keep an eye on the tide. Start at your lowest point and work your way up the shore.

- There are no highly dangerous animals on rocky shores around the British Isles, but be careful not to get spiked by a sea urchin.

Abalone

Haliotis tuberculata

Very rarely seen around the mainland coasts of Britain, nevertheless it's worth keeping an eye out for these iridescent beauties. Also known as green ormer, abalone are mottled on the outside and red or greenish-brown in colour. Meanwhile, the insides of their shells are covered in oily rainbow colours, with a slick coating of mother-of-pearl, also known as nacre. The shining interiors are hidden while the animals are alive and serve to strengthen the shells, making them extremely resilient and crack proof.

These are snails that live a similar rock-bound life to limpets, shuffling around on their foot and feeding on a diet of seaweeds. In contrast, they've evolved flatter shells, shaped like ears. They grow up to 10cm (4in) long and 6.5cm (2½in) wide, with multiple holes dotted along the back through which abalone breathe, reproduce and defecate.

Where to spot: Commonly seen around the islands of Jersey and Guernsey

I SPOTTED THIS SHELL

AT ....

ON ....

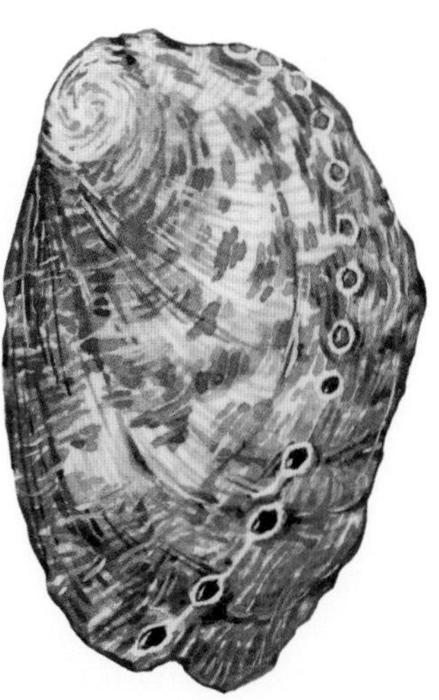

Auger

Turritellinella tricarinata

These slender shells look like spiralling augers or screws, so it shouldn't come as a surprise to learn that they push themselves a short way into sandy and muddy seabeds. White or brownish-yellow in colour, sometimes tinged with purple, these shells grow to around 3cm (1in) long and are tightly coiled – there can be as many as 20 whorls along the length.

They can live as deep as 200 metres (650ft) underwater and sometimes aggregate in large numbers. You might find lots washed up together on the shore.

Unusually for snails, augers are filter feeders, sifting small particles of food from the water – a characteristic that is more common among bivalves.

Where to spot: All around the British Isles except south-east England

I SPOTTED THIS SHELL

AT ....

ON ....

Banded wedge shell

Donax vittatus

Lying on sandy beaches, these slender, wedge-shaped shells look rather like small butterflies with their wings spread out. Their glossy, delicate shells, roughly 3cm (1in) long, can be white, purple, brown or yellow and often have a violet sheen on the inside. Delicate bands run across the outside and fine lines radiate from the point where the shells join together. Run your finger along the shells' edges and you'll feel the fine serrations. A similar-looking species is the smooth donax (*Donax variegatus*), which has edges that are finer and smoother to the touch.

Wedge shells live from the intertidal zone down to around 20 metres (65ft). They burrow themselves in sand just below the surface, which means they're often dislodged by rough seas. At low tide, watch out for the live animals at the water's edge, quickly digging themselves down into the sand using their strong, muscly foot.

Where to spot: Common on exposed sandy beaches across the British Isles, although rarer in Ireland and Scotland

I SPOTTED THIS SHELL

AT ....

ON ....

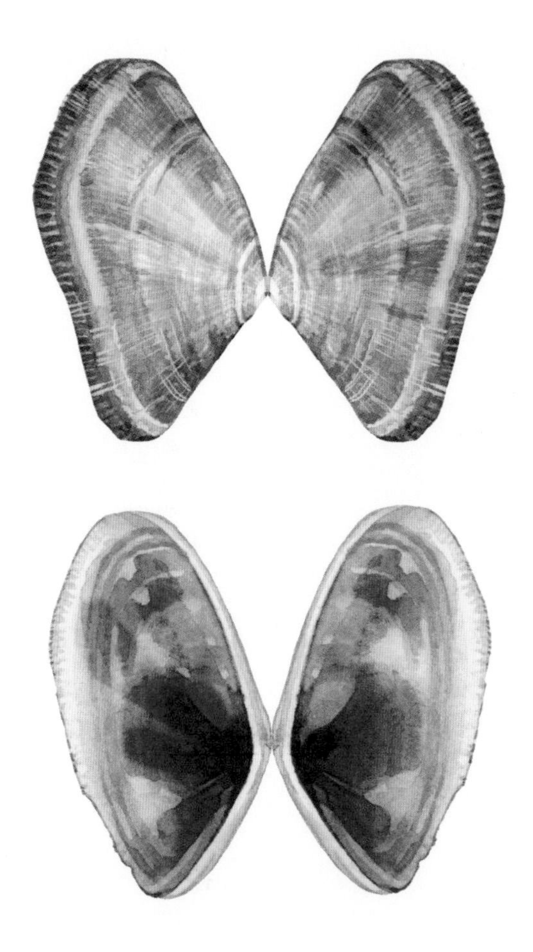

Beer barrel

Acteon tornatilis

The beer barrel has such a distinctive shell – with white bands with dark pink margins – that you should be able to spot one even from a small fragment. Intact shells are around 2–3cm (¾–1in) long and there's a distinct tooth in the lip of the opening.

These are members of the Acteonidae family, the barrel bubble snails. They live in the intertidal zone down to at least 250 metres (820ft) underwater, where they use their head with four large lobes to burrow into clean, fine sand. Like all the bubble snails, beer barrels are predators. They specialise in hunting for worms that also live buried in the seabed.

Where to spot: Fairly rare but can be seen all around the British Isles

I SPOTTED THIS SHELL

AT ....

ON ....

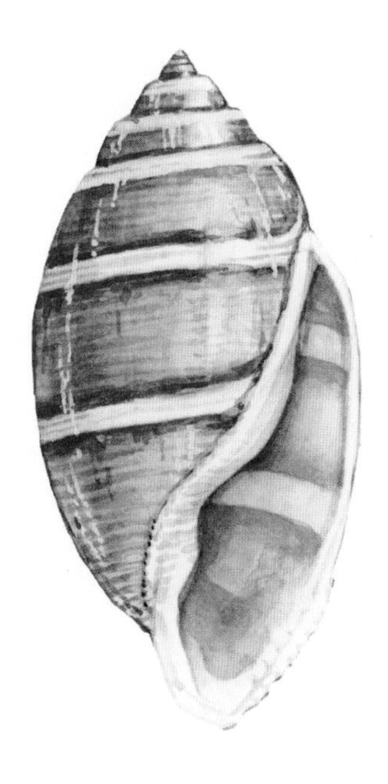

Blue mussel

Mytilus edulis

If you're keen on seafood then you're probably familiar with these blue bivalves with a pink nubbin of cooked meat inside. Blue mussels grow on intertidal rocks along exposed coasts where they withstand the pounding waves thanks to their super-strong byssus threads that anchor them firmly in place. These are the stringy beards that need yanking out before cooking. When attacked by predatory snails, such as dog whelks, mussels respond by wrapping them up in sticky byssus threads.

The shells vary in colour from purple and blue to brown and black. They have fine concentric growth lines and sometimes you'll find them with radial stripes. On the inner surface, an oval scar shows where the adductor muscle was attached, which the mussel contracts to close its shells tightly together.

Blue mussels can aggregate in huge numbers, forming dense beds. As filter feeders, they help improve water quality by removing nutrients, bacteria and toxins.

Where to spot: Common on rocky shores around the British Isles

I SPOTTED THIS SHELL

AT ....

ON ....

Blue-rayed limpet

Patella pellucida

Time your visit to a rocky shore for the lowest tides and you could be treated to the shiniest treasures of the coast. Blue-rayed limpets live on fronds of kelp that live down in the sublittoral zone of the shoreline and are only exposed briefly when the tide falls. The best way to see them is to grab a mask and snorkel and go for a swim in the shallows.

These fingernail-sized snails have iridescent turquoise stripes across their shells, which gleam when they catch the light. This may mimic brightly coloured and toxic sea slugs to discourage predators from eating them.

Material scientists have uncovered the secret to the blue-rayed limpet's dazzling colours. Minute zigzag-shaped structures in their calcium carbonate shells reflect blue and green light, while spherical particles absorb all other colours in sunlight. People are now wondering if they can imitate this effect to manufacture devices such as transparent display screens.

Where to spot: All around the British Isles except eastern coasts surrounding the Wash

I SPOTTED THIS SHELL

AT ....

ON ....

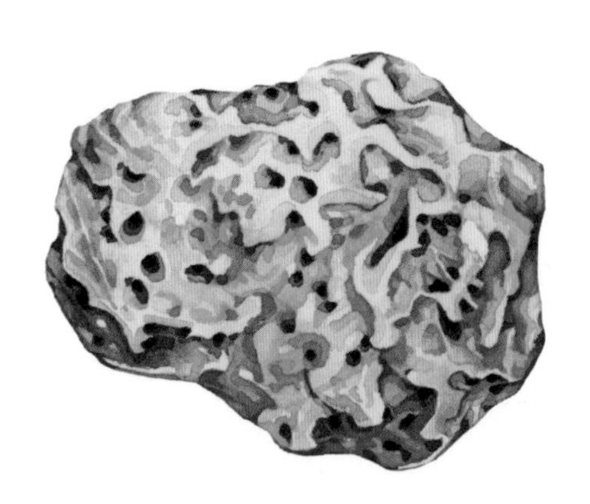

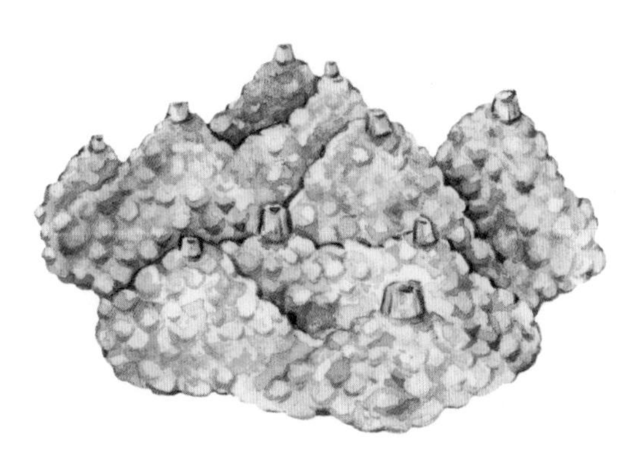

Shell Clues

As well as identifying the species of shells you find, it's also possible to find clues left behind by the animals that made them. For instance, if you find a neat, round hole punched in a shell – especially a bivalve shell – this is evidence that it was hunted and eaten by a predatory snail, such as a whelk or necklace shell.

Shells dotted with multiple holes (see opposite, top) were not eaten by numerous predators, but they were covered in a type of bright yellow sponge. Known as a boring sponge or *Cliona celata* (see opposite, bottom), it uses acid to perforate limestone rocks and the shells of living molluscs.

For some types of shell, it's also possible to make a rough estimate of how old the animal was that made it. Molluscs lay down layers of material with darker yearly bands, like tree rings. To determine an accurate age, scientists cut thin slices of shell and count the annual bands under a microscope.

Blunt gaper

Mya truncata

This is a fairly large bivalve shell, up to 7.5cm (3in) long. Unusually, the twinned shells don't press tightly together; there's a large gaping hole between them. It looks like something chopped off one end of the shells. The reason for the opening is to allow the extremely large siphon tube to stick out – it can reach four times the shell's length. Blunt gapers live buried deep in the seabed and stretch their siphons up to suck in water to breathe and filter particles of food.

They live far to the north and are widely distributed in the Arctic, where they are one of the favourite foods of walruses.

Where to spot: All around the British Isles

I SPOTTED THIS SHELL

AT ....

ON ....

Chitons

Polyplacophora

It's well worth searching for these little molluscs under rocks at low tide. They look like a cross between a slug and a woodlouse. Generally the size of a fingernail, they have shells made of eight interlocking plates arranged across their back surrounded by an oval ring of muscular tissue called the girdle. There's a dozen or so similar-looking species to find around the British Isles. They are not gastropods or bivalves, but all belong to a distinct group of molluscs called polyplacophorans.

In a similar way to limpets, chitons use their tongue-like radula to graze on films of seaweed on rocks and they creep around on their muscular foot. If they're knocked off their rock they curl up, like tiny armadillos.

Where to spot: On rocky shores all around the British Isles

I SPOTTED THIS SHELL

AT ....

ON ....

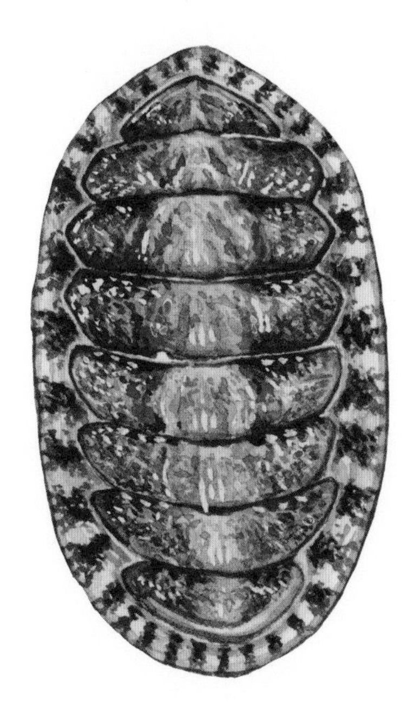

Common cockle

Cerastoderma edule

Common cockles are quintessential seashells from the seashore and ideal for decorating sandcastles. The sturdy, oval-shaped shells are covered in ribs, often grubby white in colour and generally no bigger than a 50p coin. Usually you'll find single cockles that have split apart from their twin. If you can find two that fit together you'll see that from side on they form a heart shape. In Denmark they're known as *hjertemuslinger* and in Germany *Herzmuscheln* – heart mussels.

During the winter, cockles grow slowly, or not at all, which gives their shells distinct, darker growth lines. Counting these lines, in a similar way to tree rings, gives an idea of the age of the cockle that made the shell. Cockles can live for up to seven years.

These burrowing molluscs live just under the surface of the sand and use their muscly foot to move around.

Where to spot: Common in sandy bays and estuaries across the British Isles

I SPOTTED THIS SHELL

AT ....

ON ....

Common limpet

Patella vulgata

Common limpets are a familiar sight on rocky shores, clamped tightly to rocks at low tide. Their volcano-shaped shells equip them well for life on exposed shores. Their flattened shape, up to 3cm (1in) high, makes it less likely that they'll get swept away by waves, and the large surface area of the base, 6cm (2in) across, gives them a strong anchor for their foot to cling on to. A combination of suction and sticky mucus holds them firmly in place.

When the tide comes back in, limpets set off in search of food. If you sit quietly near the waterline you might hear the rasping of their toothy tongues (the radulas) as they scrape up seaweed. Limpet teeth are some of the strongest known biological materials.

These snails have an amazing homing ability. As the tide falls, they return to the exact same patch of rock, perhaps navigating by chemical trails. Look out for the rings they etch into the rock, which help them to fix in place.

Where to spot: Common on rocky shores around the British Isles

I SPOTTED THIS SHELL

AT ....

ON ....

Common periwinkle

Littorina littorea

Common periwinkles may not be the most exciting of the sea snails, but they are widespread inhabitants on rocky shores, and easy to find nestled together in damp nooks under rocks and in rock pools at low tide. They look rather like land snails with their dark greyish-brown, banded shells, although with a more pointed apex.

These are herbivorous snails, 3–5cm (1–2in) long, that use their radula to graze the algae off rocks. At low tide, you can find them creeping around with a pair of tentacles sticking out from under their shell. Look out for the winding trails they leave behind in rock pools as they trundle through thin layers of sand.

Periwinkles have been an important food for humans for millennia and have often been found in prehistoric piles of discarded shells, called middens. These days, they are a seaside snack, served boiled and with a pin to 'winkle' them out of their shells.

Where to spot: On rocky shores all across the British Isles

I SPOTTED THIS SHELL

AT ....

ON ....

Common piddock

Pholas dactylus

Members of the piddock family, the Pholadidae, are commonly known as angelwings. A pair of these elongated, oval, white shells arranged side by side do have a rather angelic, wing-like appearance. The surface is sculpted in concentric ridges and radiating lines.

Common piddocks are boring bivalves – meaning that they bore into soft rocks, not that they're dull to be around. In fact, there's something rather dazzling about these finger-length molluscs: they glow in the dark. Bioluminescence is a rare ability among the bivalves and it's not obvious why exactly piddocks give off a dim blue-green light around their edges.

They begin boring from their larval stage. As they grow, they excavate a burrow around themselves, where they will live out the rest of their lives. Their long siphon tube, up to twice their body length, reaches up to suck in water so that they can breathe and filter feed.

Where to spot: Intertidal sandstone and chalk rocks, especially along England's southern coast and North Norfolk's chalk reef

I SPOTTED THIS SHELL

AT ....

ON ....

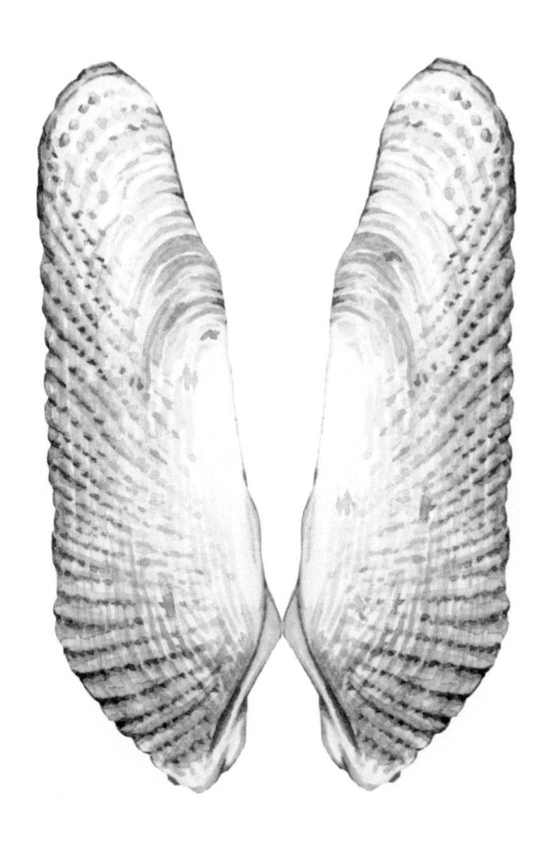

Cuttlebone

Sepia officinalis

You might perhaps mistake a cuttlebone for a piece of polystyrene foam, shaped like a little surfboard, as much as a handspan in length, washed up on the tideline. They're light, chalky and spongy, and you can easily squeeze them between your fingers and leave a mark. They're made by the common cuttlefish, a squid-like, soft-bodied mollusc and a distant cousin of gastropods and bivalves. So of course these aren't really bones (molluscs are boneless invertebrates) and they're not strictly seashells, but they did originate in a cuttlefish ancestor that had an external shell. Over time, the shell evolved into an internal structure that acts as a buoyancy device, helping cuttlefish to float underwater.

Cuttlefish live in water down to 200 metres (650ft). Studies show that these animals are intelligent, like their relatives the octopus and squid.

Where to spot: All around the British Isles

I SPOTTED THIS SHELL

AT ....

ON ....

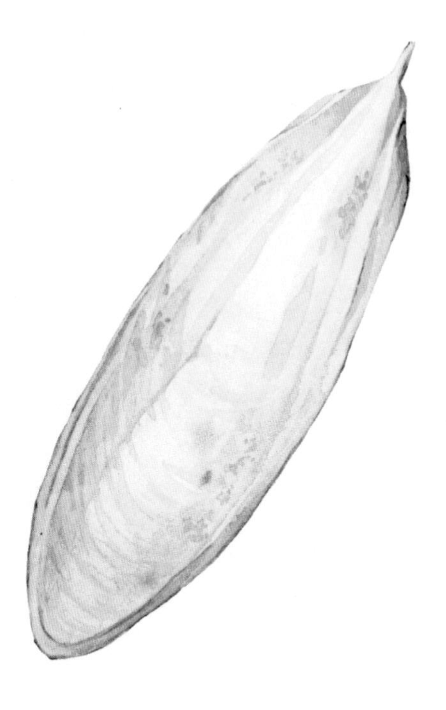

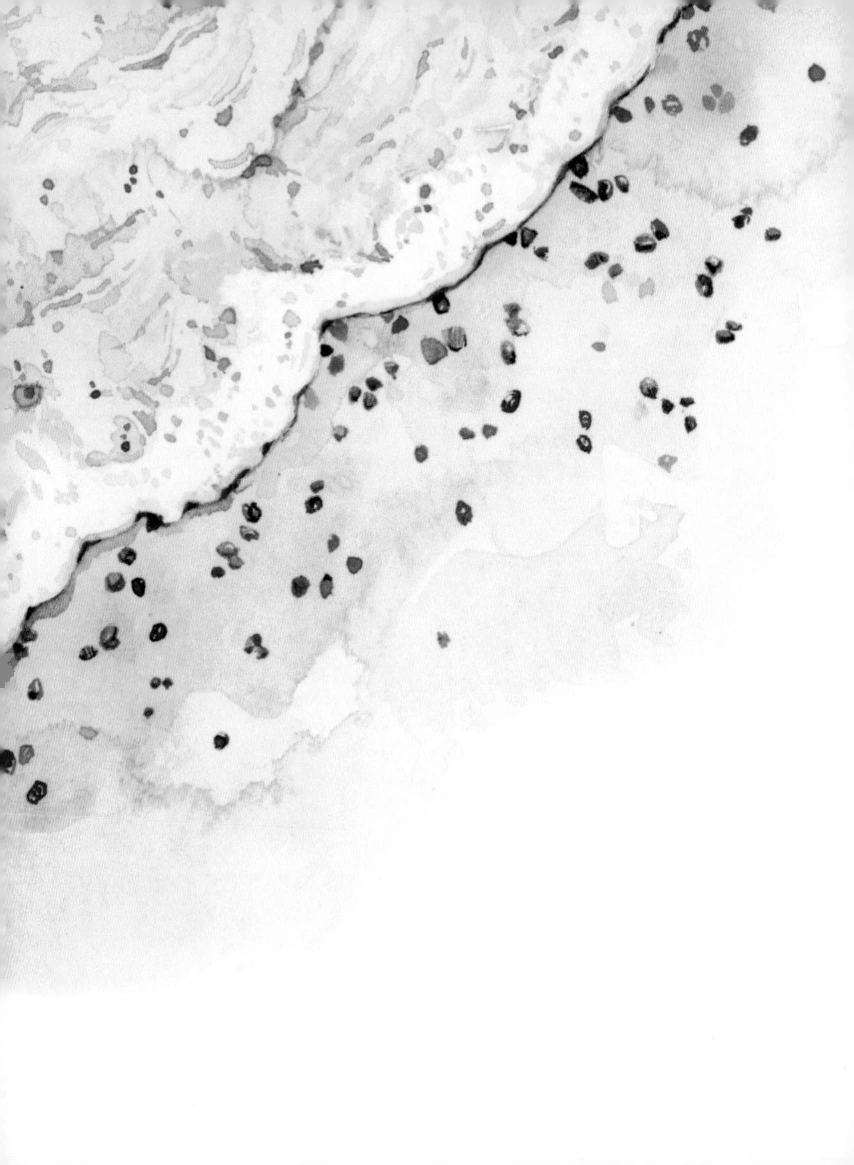

Tides

As you explore beaches and rocky shores for seashells you'll become familiar with the rising and falling of the tides. At certain times of the lunar calendar you may arrive at the coast to find the beach completely covered in water, far higher than a previous visit, or you may find much more beach exposed than before. These are spring tides, when water levels reach both their highest and lowest points. Extreme low tides are a shell spotter's delight because more oceanic animals and their shelly remains are accessible.

Spring tides happen twice a month, at new and full moons, when the Earth, Sun and Moon are all in alignment. This means the Sun's gravitational pull is added to the Moon's, causing the ocean to bulge slightly more and creating higher and also lower tides. A week after spring tides come neap tides, when the water levels are less extreme and don't reach as high or as low on the shore.

Dog whelk

Nucella lapillus

Dog whelks have spiralling shells with a pleasingly pointed spire. Around the British Isles they're usually white or yellow, but keep an eye out for other colours and patterns; they can be orange, greenish, pink, stripy, and some have black and white bands. They're usually around 3cm (1in) long (roughly thumb-sized to your knuckle) and most of the shell is made up of the final, expanded whorl.

They're predators and jam the lip of their shell between the shells of a bivalve to stop them from closing. Dog whelks also drill holes through the shells of their prey with their rasping tongue (the radula). If you find a shell with a neat circular hole punched in it, the owner was probably hunted and eaten by a dog whelk.

At low tide, look for clusters of their eggs, like swollen grains of rice, stuck to the underside of rocks.

Where to spot: Common across all rocky coasts around the British Isles

I SPOTTED THIS SHELL

AT ....

ON ....

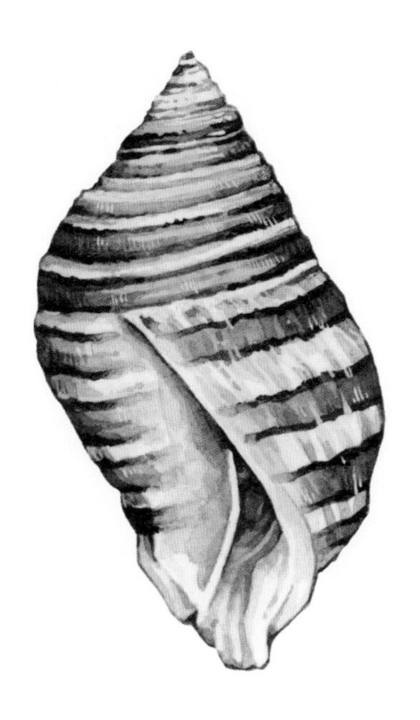

Fan mussel

Atrina fragilis

One of the biggest bivalves in the British Isles, fan mussels are also one of the rarest and hardest to find. Their triangular, golden yellow-brown shells have been known to grow up to 48cm (19in) long – a forearm's length. In the wild, fan mussels poke their tapering point into soft sediment, burying up to two thirds of their shell in the seabed. They anchor themselves in place with fine, silky byssus threads attached to small stones or shells.

Fan mussel shells are brittle and easily torn away by trawl nets and scallop dredges dragged across the seabed, which is an important reason why they are so rare. The species is protected in the UK under the Wildlife and Countryside Act. Since 1998 it's been illegal to deliberately kill, damage or sell fan mussels. If you are lucky enough to spot one, the best thing is to have a good look, take photos, then leave it be.

Where to spot: Extremely rare; mostly in deep waters off Scotland and with scattered coastal records, including Pembrokeshire and Northern Ireland

I SPOTTED THIS SHELL

AT ....

ON ....

Flat periwinkle

Littorina obtusata and *L. fabalis*

Among the seaweeds at low tide, look for flat periwinkles (*Littorina obtusata*). Their spiralling shells are the size of a fingernail, with a teardrop-shaped opening. They come in a variety of colours and patterns – bright yellow, orange, olive green, brown, banded or chequered. On more sheltered shores they tend to be lighter in colour; on more exposed shores they are darker.

These are herbivorous snails and they live in and among their seaweed food. Often you'll find them living on bladder wrack seaweeds where their colour and shape disguise them as the gas-filled bladders that help the seaweed fronds to float when the tide is in.

The sister species, *Littorina fabalis*, is very difficult to tell apart by eye (you would need to dissect them to be sure).

Where to spot: On rocky shores all across the British Isles

I SPOTTED THIS SHELL

AT ....

ON ....

Green sea urchin

Psammechinus miliaris

This is not a mollusc, but nevertheless a stunning and extremely delicate shell to find. Living sea urchins may look formidable and spiky, but their empty exoskeletons are easily crushed. Roughly the size of a squash ball that's been gently squeezed and flattened, they are greenish in colour, often with spines tipped in purple. Look carefully and you'll see that they are not entirely circular, but have a five-sided symmetry, giving away their close relationship with starfish. Green sea urchins live on rocky shores, on seagrass meadows, among seaweeds and on the seabed down to depths of 100 metres (320ft).

You'll earn shell-spotting kudos if you can find an intact sea urchin exoskeleton with its mouth parts – known as 'Aristotle's lantern' because they resemble a type of old lamp with five windowpanes – still inside. Sea urchins use this intricate apparatus, made up of five sharp teeth-like structures, to rasp at seaweeds.

Where to spot: All around the British Isles, especially in the North Sea

I SPOTTED THIS SHELL

AT ....

ON ....

Grey and purple top shells

Steromphala cineraria and *S. umbilicalis*

These charming and widespread gems are easy to find on rocky shores, but not always so easy to tell apart. Both are around the size of a 5p coin, with tightly spiralling shells and diagonal stripes, often in pinks and purples. Of the two, purple top shells, also known as flat top shells, are slightly broader, rounder and flatter. On the underside they have a deep, round hole, called the umbilicus (which is where their species name comes from). Grey top shells have an oval-shaped hole and their shells are more conical in shape.

The outer layer of the shells of both species can wear away to reveal the lustrous, shiny mother-of-pearl (or nacre) underneath.

Keep an eye out for them in rock pools, roaming around searching for seaweeds to graze on, with a pair of stripy antennae sticking out from their shells.

Where to spot: On rocky coastlines all around the British Isles, though purple top shells are rare on eastern coasts of England and Ireland

I SPOTTED THIS SHELL

AT ....

ON ....

Horse mussel

Modiolus modiolus

Horse mussels look similar to blue mussels but are bigger, up to 20cm (8in), and covered in a shiny brown periostracum. They can live singly but more interesting things happen when they cluster in dense beds. Both living and empty shells cover the seabed, knitted together with byssus threads, creating sheltered nooks for an assortment of other species: crabs, sea urchins, sea anemones, scallops, brittle stars, feather stars, barnacles, tube worms, soft corals, sea squirts, whelks and sponges.

Most of the UK's known horse mussel beds are in Scotland. The biggest is Noss Head in Shetland, covering almost 4 square kilometres (1.5 square miles).

Horse mussel beds are easily destroyed by bottom trawling and scallop dredging, and are slow to recover. They're recognised as a Priority Marine Feature in Scotland and a UK Biodiversity Action Plan Habitat, and are safeguarded inside marine protected areas.

Where to spot: All around the British Isles, with most of the large, dense beds off Britain's northern and western coasts, generally 20–50m (65–160ft) down

I SPOTTED THIS SHELL

AT ....

ON ....

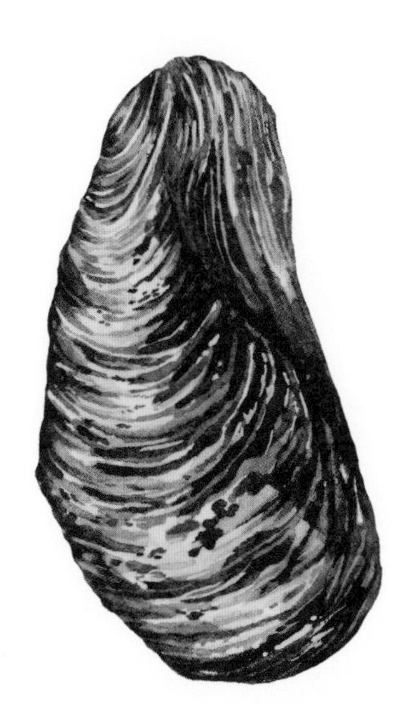

Icelandic cyprine

Arctica islandica

The greatest claim to fame for these rather nondescript clams is that they can live for centuries. A specimen dredged off the coast of Iceland in 2006 was a record-breaker. Painstakingly counting the growth lines of a thin slice of shell under a microscope, scientists worked out that the clam had lived for 507 years. They named it Hafrún, a woman's name meaning 'the mystery of the ocean' (although it was not possible to work out the sex of the clam). This was the oldest known individual animal to have had its age accurately measured. Icelandic cyprines spend their long, slow lives sitting partly buried in sand, filtering the water for particles of food.

These palm-sized clams go by various other names, including mahogany clam and black clam because they have a glossy, dark periostracum covering over their white or pale brown shell.

Where to spot: Around the British Isles, although rare off the coasts of East Anglia and Ireland

I SPOTTED THIS SHELL

AT ....

ON ....

Shell Magic

People have long been enchanted by seashells. The oldest known examples of jewellery are beads fashioned from shells dating back around 150,000 years. Many cultures use shells as decorations and symbols of life and death, sex and fertility. People around the world have buried their loved ones and leaders in graves filled with precious shells. The enduring appeal of shells comes down to their pleasing, sculpted shapes, the clean white colour many of them have and the fact that they come from the hidden, mysterious realm of the ocean.

Today, myths still surround seashells, including the idea that they capture the sound of the sea. Physicists explain how this happens – that ambient sounds reverberate and resonate through the hollow interior, imitating the whoosh of waves.

Keyhole limpet

Diodora graeca

If you find a limpet that seems to have had the point of its shell cut off, take a closer look and check if in fact it's meant to be that way. Keyhole limpets evolved a neat hole in the top of their pointed, volcano-shaped shells through which they breathe water and release faeces. The hole is offset towards the front end of the oval-shaped shell (this is the end where you might spy a head poking out in living specimens).

The keyhole limpet can grow to 4cm (1½in) and its base is slightly bent – not flat like that of the common limpet. Live keyhole limpets have bright red or orange bodies, sometimes white with darker spots. They live under stones and on rocks from the lower parts of the shore, near the low water mark, and all the way down to 250 metres (820ft) underwater.

Where to spot: Widespread on coasts of Ireland and western Britain

I SPOTTED THIS SHELL

AT ....

ON ....

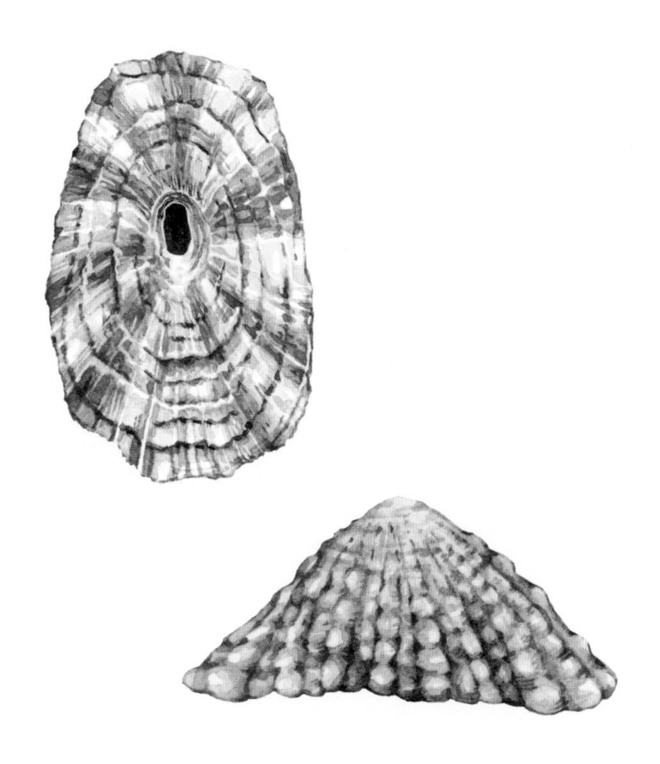

Necklace shell

Euspira catena

Look out for these smooth, polished, helical shells that are easy to spot when they get washed up on sandy beaches. Notice the delicate row of red markings that spiral inwards along the final whorl. These yellow or buff-coloured snails grow to around 3cm (1in) in size and live just below the surface of the sandy seabed. Turn them over and you'll see a distinct, deep umbilicus hole on the underside.

Keep an eye out for their distinctive egg cases that also wash up on beaches. They're known as sand collars because they look like an old-fashioned detachable collar for a shirt or blouse. Females make these structures by gluing together sand grains and then laying thousands of egg capsules in the matrix.

These predatory snails hunt for other molluscs. They drill into bivalves and then suck out their soft insides, leaving a telltale neat round hole.

Where to spot: All around the British Isles

I SPOTTED THIS SHELL

AT ....

ON ....

Netted dog whelk

Tritia reticulata

The distinctive criss-cross patterning on this shell makes it look as if the snail was pressed into a net. Netted dog whelks inhabit the lower parts of rocky shores where live specimens can sometimes be seen out and about hunting for food, with their siphon sweeping the water to sniff for dead and dying prey – these are scavenging snails. The shells grow to 3cm (1in) and have an oval aperture, lined with little teeth and with a short canal that the siphon pokes through. Look out for the clear egg capsules with tiny yellow eggs inside that females lay on seagrasses, seaweeds and under rocks.

Their name is deceiving because these are not in fact related to common whelks, but belong to a different family, the Nassariidae, known as mud snails.

Where to spot: On most coasts around the British Isles, except East Anglia and north-east Scotland

I SPOTTED THIS SHELL

AT ....

ON ....

Painted top shell

Calliostoma zizyphinum

In my humble opinion, this is the most beautiful British seashell. Get yourself down to the lowest parts of a rocky shore, especially during spring tides, and you stand a good chance of finding these gastropods hiding among seaweeds.

Their conical, straight-sided, 3cm (1in) tall shells really do look like old-fashioned spinning tops, and they come painted in bright pink and purple patterns. You might occasionally find a completely white one. A living top shell keeps its shell clean by wiping over it with its extendable foot. Older shells that have been worn away by waves and sand will show their shining mother-of-pearl underneath.

If you're lucky enough to see a living one, watch quietly and see if you can catch sight of its body, which is just as colourful as its shell and flecked with purple, crimson, burgundy or yellow.

Where to spot: On rocky shores all around the British Isles

I SPOTTED THIS SHELL

AT ....

ON ....

Pelican's foot

Aporrhais pespelecani

More than 250 years ago, Swedish naturalist Carl Linnaeus gave the pelican's foot its charmingly appropriate name in his landmark book of species, *Systema Naturae.* These shells clearly resemble the splayed webbed foot of a large waterbird. Pelicans use their feet for paddling, and for the snails the webbed shape helps them sit their shells on the surface of the sandy or muddy seabed without sinking in, like a snowshoe. The spikes presumably also help make the shell more unwieldy for predators to handle.

An intact pelican's foot shell is well worth treasuring. The snails live in the sublittoral zone and their shells only wash up on beaches occasionally. The 'toes' often break off and so damaged shells are a more common find. When searching for broken specimens, look out for the rounded knobs on the tall, tapered spire. The 5cm (2in) shells are glossy and creamy or sandy in colour, often with a purple tint.

Where to spot: Potentially anywhere around the British Isles

I SPOTTED THIS SHELL

AT ....

ON ....

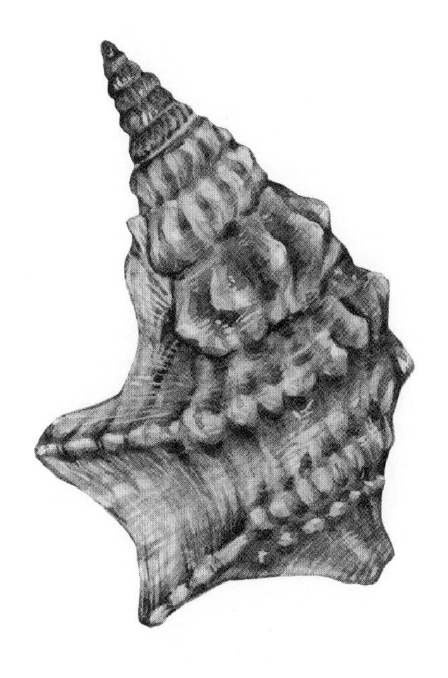

Queen and king scallops

Aequipecten opercularis and *Pecten maximus*

These are some of the largest and most striking bivalves to spot around British coasts, with their ridged, fan-shaped shells that easily grow to the size of your palm. Kings and queens can both be quite colourful, with pink and orange mottled shells. You can tell them apart by the shape of the 'ears' sticking out either side: in king scallops (shown opposite) these are the same shape; in queen scallops (pp. 2 and 7) the side projections are asymmetrical, with one bigger than the other. King scallops also have one curved shell that lies in small self-dug hollows in the seabed, with a flatter shell on top.

Scallops are filter feeders and they have hundreds of small, gleaming eyes dotted around their mantle which can detect light and dark. They are hunted by starfish and escape by squeezing their shells together and squirting jets of water through their siphon tube.

Where to spot: All around the British Isles; king scallops are rarer along the east coasts of Scotland and England

I SPOTTED THIS SHELL

AT ....

ON ....

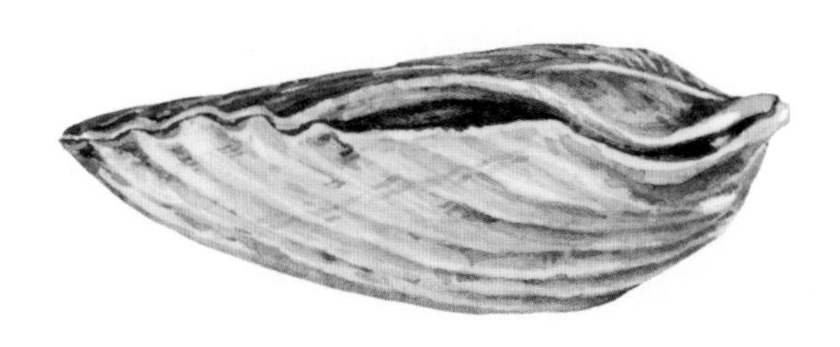

Rayed artemis

Dosinia exoleta

These large and pleasingly circular bivalve shells fit comfortably in the palm of your hand. Two shells in a pair match each other in shape and size. They are generally white, yellowish or light brown, with fine concentric ridges running across them. The deeper ridges are annual growth lines; count them and it will give you an estimate of the shell's age (which can be as much as 10 years old).

Their patterning is highly variable. The most splendid specimens are covered in pink and brown patterns of intricate zigzags, radiating stripes or triangles. They look like someone with an eye for mathematically derived patterns has scribbled all over them.

Pigments are laid down as the shell grows and scientists are not entirely sure how this happens, or for what purpose – if any – the patterns serve the bivalve. This species lives burrowed in muddy and shelly gravel, so their beautiful patterns often go unseen.

Where to spot: Common around the British Isles, although not in the south-east of England

I SPOTTED THIS SHELL

AT ....

ON ....

Razor clam

Ensis ensis

There's no mistaking the group of molluscs called razor clams – also known as sword razors, or spoots in Scotland. At around 13cm (5in) long, they look like overgrown toenails with a slight curve. Their paired shells are often covered in a flaky, olive-green layer of periostracum. Their inner surfaces are white with a purple sheen. You'll find them on sandy beaches, such as the Brancaster Estate bordering the Wash on the North Norfolk coast. Sometimes they wash up in great piles, which crunch when you walk over them – these shells are surprisingly fragile.

In life, the clams live buried vertically in the sand and leave their pair of siphon tubes sticking above the seabed, inhaling and exhaling seawater to breathe and filter tiny particles of food. At low tide, look out for the key-shaped dimples in the sand, which show where razor clams are buried.

There are two other species of razor clam – *Ensis siliqua* and *E. magnus* – which have slightly longer and straighter shells.

Where to spot: Common on sandy beaches across the British Isles

I SPOTTED THIS SHELL

AT ....

ON ....

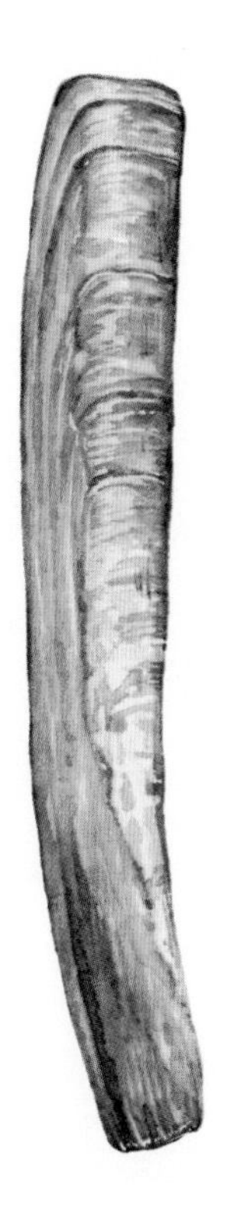
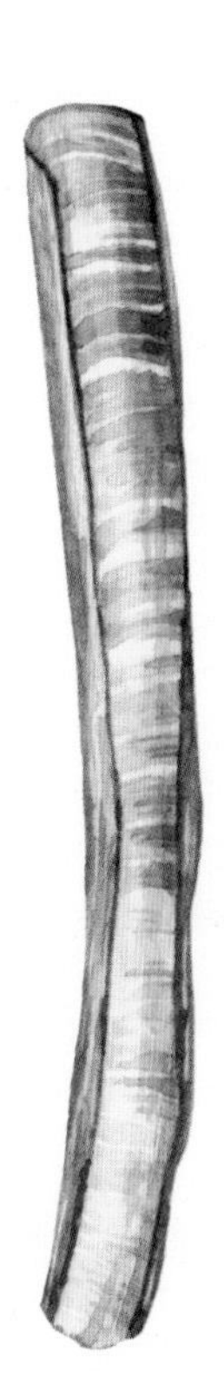

The Art of Shells

For hundreds of years humans have used shells to make arts and crafts, from chambered nautilus shells carved into ornate, gilded cups, to shell grottos such as A la Ronde, a quirky 16-sided National Trust house in Devon. But best, and most mysterious, are the shells' own artworks. Many seashells are covered in patterns of stripes, zigzags, triangles and spots. Tropical shells, such as cone shells, tend to be the most flamboyant, but shells around the British Isles also have delicate patterns to look out for. Molluscs lay down pigments in their shells as they grow. For some species the colours help to camouflage them; this is the case for flat periwinkles, which look like gas-filled bladders on seaweeds. But in many cases it's not obvious why molluscs go to the trouble of decorating their shells – especially when they spend much of their lives buried in the seabed, out of sight. One possible explanation for why they do this could come down to the stop-start nature of shell making. The patterns could simply be the mollusc's way of reminding itself where to line up its mantle so that the next time it makes more shell it carries on in the right orientation – or it could just be a way of building a neater shell.

Rough periwinkle

Littorina saxatilis

Small, plump, spiralling shells with deep, neat ridges, you'll find rough periwinkles hunkered in crevices between rocks. Also look out for them nestled inside empty barnacle shells. They're small enough to fit neatly on a 5p coin.

Probably the greatest claim to fame for these little snails is that they've been wrongly named at least a hundred times – likely more than any other species on the planet. One reason is that periwinkles were a favourite of Victorian naturalists who busily rooted around the rocky shores of Britain, eager to discover new species. Dozens of species recorded in the 1800s were later deemed to be unscientific splitting of the single species. And to be fair, rough periwinkles exist in populations that vary slightly in appearance from place to place – some have thicker or thinner shells, some have a bigger foot. Scientific discussions continue as to whether these should be considered as subspecies.

Where to spot: On rocky shores all across the British Isles, as well as salt marshes and seagrass meadows

I SPOTTED THIS SHELL

AT ....

ON ....

Saddle oyster

Anomia ephippium

Look out for the gleaming sheen of these super-shiny shells on beaches and among seaweeds. They are brittle and bumpy, growing to fit the shape of the surface they live on, be that a stone or another shell. Also known as jingle shells, saddle oysters hold themselves in place on hard surfaces with a bunch of strong byssus threads, which poke out through a hole in the lower part of the shell. Young saddle oyster spat look like tiny painted fingernails fixed to the underside of rocks.

They are transparent and can be tinged in yellow, grey, blue and orange, and are roughly circular, up to 6cm (2in) across.

Where to spot: All around the British Isles

I SPOTTED THIS SHELL

AT ....

ON ....

Slipper limpet

Crepidula fornicata

These well-named snails do indeed look like slippers. They have a rounded upper side and a shelf underneath that covers half their length, leaving space to slide in a little foot. The shells are smooth, pale pink or yellowish, with darker patterns of dots and dashes. They aren't in fact limpets and they're not native to the British Isles, but were likely brought over when oysters were imported from North America in the nineteenth century.

Slipper limpets have unusual mating habits (their scientific species name is a hint). Look out for piles of slipper limpets, stacked up one on top of the other. The largest snail, 5cm (2in), at the bottom of the pile is a female; the smallest at the top are males; the ones in between are in the process of changing sex from male to female. Males mate with the female and she broods her eggs in her shell before they hatch into larvae and swim off.

Where to spot: Common across southern coasts of England and Wales

I SPOTTED THIS SHELL

AT ....

ON ....

Spiny cockle

Acanthocardia aculeata

Spiny cockles look like large versions of common cockles that are covered in sharp prickles. They can be the size of the palm of your hand. If you're lucky enough to find a matching pair, you'll see they have a similar heart-shaped profile to common cockles.

The spines are no doubt a useful defence, discouraging predators from trying to eat them. They may also help the cockles lodge themselves in the seabed and prevent currents winnowing the sand from around them.

The shells are pale brown on the outside and white on the inside. Unlike common cockles, which have a mostly smooth interior, spiny cockles have corrugated ribs that extend through to the inside of the shells.

Live spiny cockles have a bright red foot, which they poke out between their shells like a long tongue and use to push themselves along. They live below the low tidemark, down to at least 50 metres (160ft), and occasionally their shells wash up on beaches.

Where to spot: Occasionally wash up around the British Isles

I SPOTTED THIS SHELL

AT ....

ON ....

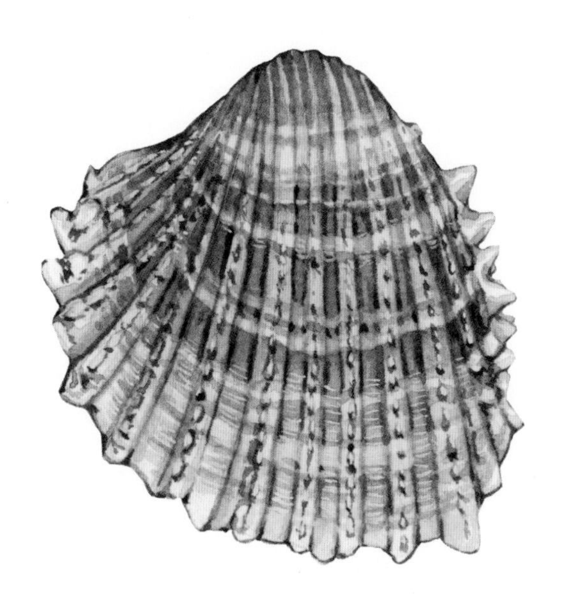

Spirorbis tube worm

Spirorbis spp.

Take a close look at various types of seaweed growing lower down the shore towards the low water mark, such as serrated wrack (*Fucus serratus*) and bladder wrack (*Fucus vesiculosus*), and you might spy what look like hundreds of tiny white coiling shells. They also live stuck to the underside of rocks and stones. But don't be deceived! These aren't molluscs but a type of worm that builds the smooth white tube in which it lives.

The worms are a few millimetres long with a bright orange body. They poke a crown of tentacles out of their tubes to filter food particles from the seawater when the tide is in.

Where to spot: On rocky shores all around the British Isles, especially western Scotland

I SPOTTED THIS SHELL

AT ....

ON ....

Spotted cowrie and Arctic cowrie

Trivia monacha and *T. arctica*

These are tiny, egg-shaped seashells, at most the size of a fingernail, which makes them all the more rewarding to find. If you visit a rocky shore at low tide, be sure to search carefully under rocks, in crevices and little caves, and you might spot living cowries, hanging upside-down from their bright orange foot. A good way to spot them is to look for their favourite food. If you can find sea squirts, which grow in colonies on the rocks and sometimes look like little flowers, then you may well find cowries sniffing around nearby with their orange siphon.

Like other cowries, these lap their soft tissue, called the mantle, over the top of their shells, helping to keep them shining like fine porcelain. These two European species are a similar size. You can tell them apart by the presence – or otherwise – of spots: adult Arctic cowries don't have any.

Where to spot: Mainly on western and northern coasts of Britain and Ireland

I SPOTTED THIS SHELL

AT ....

ON ....

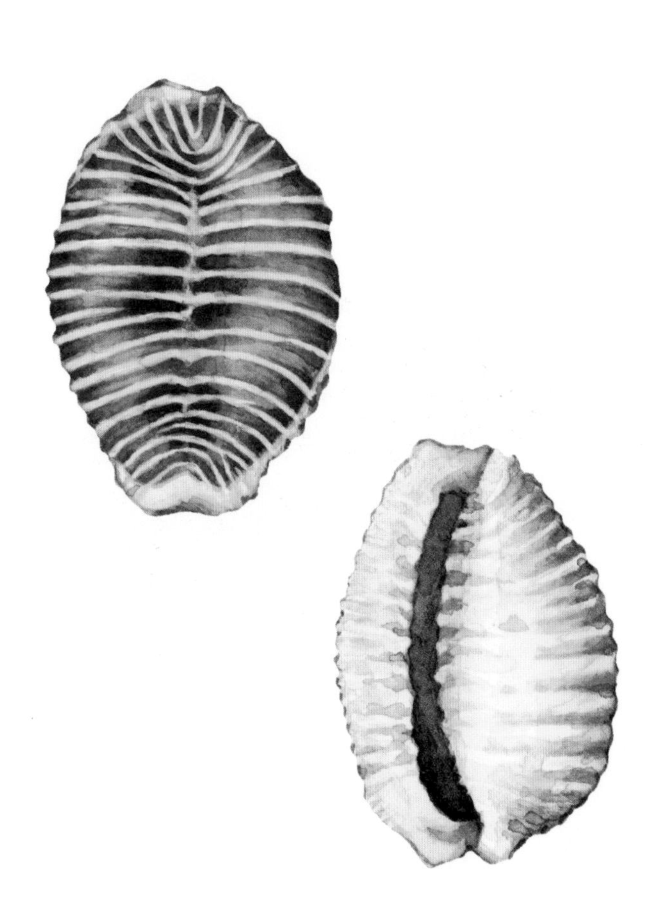

Shell Borrowers

Humans aren't the only animals that like collecting shells. Hermit crabs lost their ability to make a protective exoskeleton and instead evolved to live inside empty seashells. Hundreds of species of hermit crabs live in the ocean and several are native to the British Isles. If you spot a seashell that isn't gliding smoothly but scuttling along in a tide pool, then you've likely found a hermit crab. Turn it over carefully and you might spot a pair of claws and antennae sticking out. Some hermits evolved one enlarged claw that neatly seals the open hole, like a door.

When a hermit crab gets too big for its shell it needs to find a new one. Sometimes, hermits gather together and form orderly queues – the biggest at one end, smallest at the other. Then, in a flurry of shells, they all swap with their neighbours.

Spotted sea hare

Aplysia punctata

Sea hares are a type of snail that keep their shells on the inside. Translucent and amber coloured, these look like one half of a mussel shell. They are thinner and smaller than many other snail shells – the largest grow to 6cm (2in) inside 20cm (8in) sea hares – but that's enough to help protect the sea hare's heart and other internal organs.

Being so flimsy, these shells aren't easy to find intact and you'll probably have more luck spotting living sea hares in rock pools. They get their name from the pair of long tentacles on their head, which look rather like ears.

Like their terrestrial namesakes, sea hares are herbivores. The type of seaweed they eat seems to give them their colour. Most are purplish-red, but sometimes they're brown or green.

Sea hares are simultaneous hermaphrodites – both male and female at the same time. They often cluster together and mate in chains.

Where to spot: On rocky shores and seaweeds all around the British Isles

I SPOTTED THIS SHELL

AT ....

ON ....

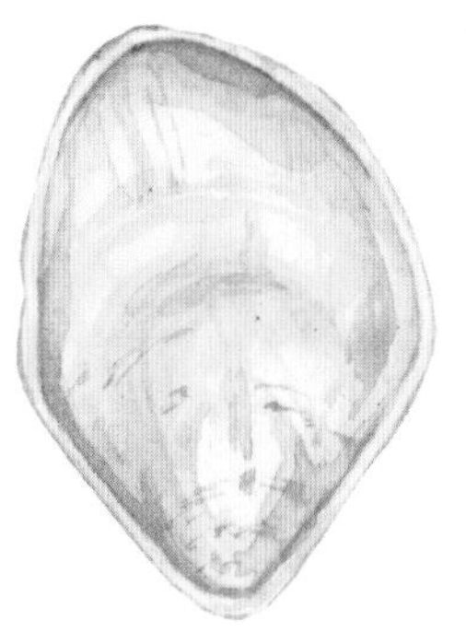

Sting winkle

Ocenebra erinaceus

Sting winkles, also known as oyster drills, are an unmistakeable species of whelk with jagged-looking shells. Malacologists describe them as being tumid (meaning swollen) and angulate (sit one on your palm and the spire angles upwards), with an outer lip that is crenulate, or finely scalloped, in young shells, and in older shells is thick and denticulate, or covered in little teeth.

The shells grow to be 5cm (2in) long and 2.5cm (1in) wide, and are white or yellowish with brown markings, especially on the ribs and ridges. They can be seen at low tide on the lower parts of sheltered rocky shores.

These predators drill into young oysters and they are sometimes considered to be pests on oyster beds. This is a native species to the British Isles. A similar-looking species of oyster drill from the same family, *Urosalpinx cinerea*, was introduced from the western Atlantic and can now be found on the Kent and Essex coasts.

Where to spot: Rocky shores in Ireland and south-west Britain

I SPOTTED THIS SHELL

AT ....

ON ....

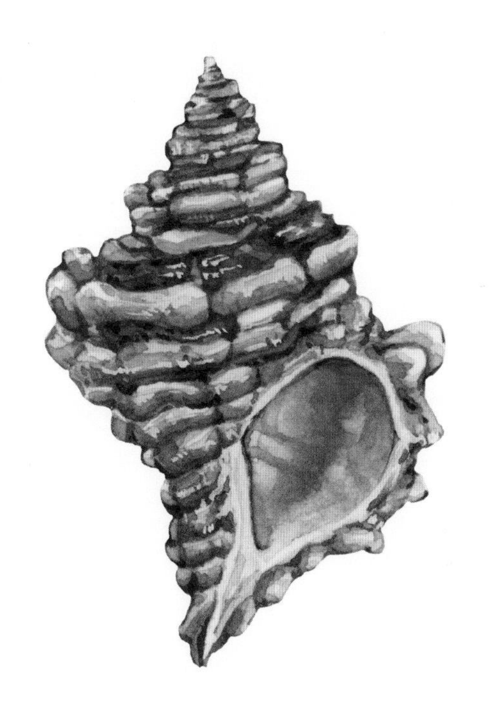

Turban top shell

Gibbula magus

Probably the biggest top shell you'll find in the British Isles and quite distinct from painted top shells, which are far more neatly conical. These do indeed look like little coiled turbans. They are low, flat and blunt, around 3cm (1in) wide. From side-on you can see they have deep steps between the main whorls, each with rounded bumps on top.

Turban top shells vary greatly in colour. They can be white, yellowish or pale brown with zigzag radiating stripes or flecks that make them look mottled all over. Some are pink and white, like raspberry ripple ice cream.

On the underside there's a deep hole (the umbilicus) with a spiral-shaped interior. Living turban top shells have a circular plate (the operculum), which acts like a trapdoor and closes the shell opening when the snail pulls its body inside, to avoid predators and drying out at low tide.

Where to spot: On rocky and shingle beaches on Britain's southern and western shores; dotted around the coast of Ireland

I SPOTTED THIS SHELL

AT ……………………………………………

ON ……………………………………………

Violet sea snail

Janthina janthina

This is a seashell I dream of finding one day, not only for its delicate, blue-violet shell but also so that I can see the home of a truly remarkable animal. Violet sea snails don't crawl around on the seabed but float at the ocean surface, hanging down from a raft of frothy bubbles made with sticky slime secreted from their foot. They belong to a unique floating ecosystem called the neuston, which contains many blue-coloured animals that match their oceanic surroundings; these include jellyfish-like animals called blue buttons and by-the-wind sailors, which the violet sea snails feed on.

Violet sea snails are members of the wentletrap family: snails with intricate spiralling shells that specialise in hunting corals and sea anemones. These small 4cm (1½in) high, 3cm (1in) wide shells float around tropical and temperate waters and occasionally drift on stormy seas to our coasts, but you'll have to be extremely lucky to find one.

Where to spot: More likely on western coasts of the British Isles bathed in the Gulf Stream

I SPOTTED THIS SHELL

AT ....

ON ....

Waved whelk

Buccinum undatum

At up to 10cm (4in) long and 6cm (2in) wide – a good handful – waved whelks (also known as common whelks) are the largest gastropod shells you're likely to find on British beaches. The shells have seven or eight whorls and are sculpted into wavy folds (the species name *undatum* means wavy). Rarely seen alive in the intertidal, they usually inhabit deeper water.

Waved whelks are classic predatory snails with a deep notch in the shell where the siphon sticks out to sniff for prey. They're members of a different family from the dog whelks but have a similar hunting strategy. A whelk will jam the lip of its shell between the shells of a bivalve, then slurp out the soft insides.

Female whelks lay spongy clumps of lentil-shaped eggs. On the strand line, look out for their empty, papery egg masses, which are known as egg clouds or sea wash balls.

Where to spot: Common all around the British Isles

I SPOTTED THIS SHELL

AT ....

ON ....

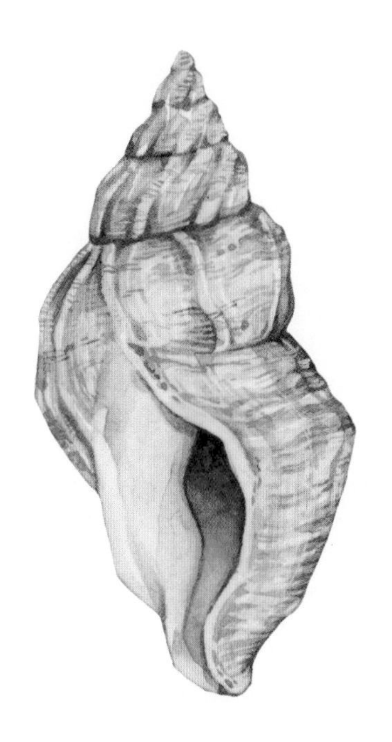

Wentletrap

Epitonium clathrus

Eighteenth-century shell collectors paid sky-high prices for specimens of precious wentletrap shells – especially *Epitonium scalare*, a species from tropical shores, which at the time was incredibly rare.

Epitonium clathrus, a smaller species from European seas, has never been worth a fortune but it is undoubtedly a beautiful little shell and a rare find that's definitely worth celebrating.

No other snail shells are quite like wentletraps with their characteristic spiral ribs, known as costae, which link together the 15 loosely wound whorls. They grow to around 4cm (1½in) long and can be creamy or shiny white with brown spots and sometimes purple bands.

Wentletraps are sublittoral species that live below the low water mark on sandy and muddy seabeds. In spring they migrate closer to shore to spawn. They are predators and feed on sea anemones.

Where to spot: Britain's southern and western shores; rare in Scotland and Ireland

I SPOTTED THIS SHELL

AT ....

ON ....

Glossary

adductor muscle	Muscle inside a bivalve that pulls the two shells together and holds them tightly closed.
angulate	Describes an object or body bent at an angle.
bioluminescence	Light emitted by living organisms either from glowing bacteria living inside their bodies, or by combining two chemicals produced in specific parts of the body.
bivalves	Molluscs with an exoskeleton divided into a pair of shells.
byssus threads	Tough, stretchy threads used by bivalves, such as mussels, to attach themselves to rocks.
cephalopods	Molluscs including octopuses, cuttlefish and squid that generally don't have an exoskeleton.
chitons	Molluscs with shells made of eight plates arranged across their backs.
clams	Widely used common name for many different kinds of bivalves, often edible varieties that live in soft sediments.
cockles	Bivalves from the family Cardiidae, with ridged shells, that are common on sandy beaches.
costae	Vertical ridges on the outside of a seashell.
crenulate	With a finely scalloped or notched edge.
denticulate	Covered in fine tooth-shaped structures.

egg clouds (or sea wash balls)	Clusters of egg capsules laid by whelk snails that float once the eggs have hatched and commonly wash up on beaches.
exoskeletons	The hard outer body covering on many animals that don't have a backbone, including molluscs.
filter feeders	Animals, including bivalves, that sift small particles of food from water, often using structures on their gills.
gastropods	Group of molluscs including slugs and snails.
girdle	Strong and flexible part of the body of a chiton that circles the shell plates and holds them together.
high water mark	The level on a shore reached by the sea at high tide. The level changes depending on the type of tide and is higher up the shore during spring tides than during neap tides.
intertidal	The space on the seashore between the high and low water marks.
larva	Early stage in the life of many marine animals, including molluscs. Usually they are small, transparent and drift through the water.
limpets	Gastropods from the Patellidae family with conical, non-spiralling shells that clamp tightly to rocks.
lower shores	Parts of the seashore nearer to the low water mark than the high water mark.
low water mark	The level on a shore reached by the sea at low tide. The level changes depending on the type of tide and is lower down the shore during spring tides than during neap tides.

malacologist	A person who studies molluscs.
mantle	The soft tissue covering the body of a mollusc that secretes its shell.
mollusc	A group of soft-bodied, spineless animals including snails, clams and octopuses.
mussels	Bivalves from the family Mytilidae that generally have elongated, dark shells that fix to rocks with byssus threads.
nacre	The shiny layer on the inside of many mollusc shells which makes the shell very tough and strong, also known as mother-of-pearl.
neap tide	When the moon is half full and the high and low tides don't move as far up and down the shoreline.
neuston	A floating ecosystem that exists at the surface of the ocean.
operculum	A hard, often circular plate secreted by gastropod molluscs that closes the opening of the shell and protects the animal inside from dehydration and predation.
periostracum	A thin organic layer on the outside of a mollusc shell that forms as the shell is made and helps to protect the shell from damage.
periwinkles	Gastropods from the Littorinidae family with small, spiralling shells.
piddocks	Bivalves from the Pholadidae family that use their elongated shells to grind holes in soft rocks and clay.
radula	Hard, tongue-like structure inside the mouth of a snail or cephalopod that is used for

feeding, such as scraping up seaweeds and chewing prey.

sea slugs Gastropods that don't produce an external shell.

sea squirt A spineless marine animal that grows on the seabed and on rocks, usually with a transparent body.

siphon Fleshy, bendy tube that molluscs use to inhale and exhale water for breathing, feeding and detecting prey. Gastropods have a single one; bivalves have a pair.

spat Stage of a bivalve's life cycle after the drifting larva has permanently settled onto a hard surface, such as a rock.

spring tides Tides around new and full moons when the high tides are at their highest and low tides are at their lowest, due to the Earth, Sun and Moon all being lined up.

strand line Part of the shore where the high tide leaves a line of debris such as seaweeds and driftwood.

sublittoral Part of the shore, below the low water mark, which is always submerged in sea.

top shells (or top snails) Gastropods from the Trochidae family with spiralling shells shaped like a toy spinning top.

tumid Having a swollen appearance.

umbilicus The hole on the underside of a snail shell at the centre of the spiral.

umbo The highest point on each shell of a bivalve.

whelks Gastropods from various groups including the true whelk family, the Buccinidae, and dog whelks, which belong to the Muricidae family.

Recommended guides

Collins Complete Guide to British Coastal Wildlife by Paul Sterry and Andrew Cleave (HarperCollins, 2012)

Shell Life on the Seashore by Philip Street (Faber & Faber, 2019; first published 1961)

The Sound of the Sea: Seashells and the Fate of the Oceans by Cynthia Barnett (W.W. Norton & Co, 2021)

Spirals in Time: The Secret Life and Curious Afterlife of Seashells by Helen Scales (Bloomsbury Publishing, 2015)

Acknowledgements

I was brought up with the National Trust deeply embedded in my life, as my family paid regular visits to the coastline, beaches and fine country houses of Cornwall, and so it's a great honour all these years later to collaborate on this book of coastal treasures. My thanks to everyone at HarperCollins and the National Trust who has helped make this happen, especially David Salmo and Peter Taylor, and my gratitude to Ella Sienna for depicting these shells so beautifully in her artwork. My love to my family for bringing me to the sea from before I can remember, and to my husband, Ivan, for being my steadfast shell-spotting companion.

Index

Note: page numbers in **bold** refer to illustrations.